The Alphabet Kids Go to the Planetarium

Based on a story by Isaac J. Kassin

Written by **Isaac J. Kassin** with **Patrice Samara**
Illustrated by **Mickie Quinn Boggs**

Meet the Alphabet Kids: Allegra, Elena, Isaac, Oni, Umar, and Yang.

They call themselves the Alphabet Kids because that's where they like to hang out - at the Alphabet Afterschool Center.

Every day they learn something new.

Today, the Alphabet Kids were taking a field trip to the planetarium.

A planetarium is a place where kids and adults can go to see stars and planets projected on the inside of a dome.

The Alphabet Kids got on the bus and Isaac's mom, Yang's dad, and Elena's dad were coming, too.

On the bus, everyone was talking about what it would be like to go to outer space. After a short drive, they arrived at the planetarium and were welcomed by a very friendly man.

"Hello everyone! My name is Dr. Blake. I am an astronomer. Does anyone know what an astronomer does?"

"No," chimed the Alphabet Kids.

"An astronomer is a scientist who studies galaxies, planets, stars, and other objects in space," answered Dr. Blake. "Today I will show you our Solar System!"

Dr. Blake led the Alphabet Kids into a very dark, round room.

"The ceiling looks just like the sky at night!" exclaimed Isaac as stars appeared on the ceiling.

"Wow! This is beautiful. Look how many stars there are in the sky." whispered Oni.

"There are billions of them," answered Elena. "*Que bonito*, it's so beautiful."

Dr. Blake explained, "Here you can see a galaxy."

"What is a galaxy?" asked Umar.

"A galaxy is a collection of many stars like the ones you see on the ceiling," answered Dr. Blake, pointing at the beautiful, sparkling stars.

"Are we in a galaxy?" asked Isaac.

"Yes," answered Dr. Blake, "The galaxy that our Solar System is in is called The Milky Way."

Elena raised her hand and asked, "What is a Solar System?"

Dr. Blake showed a picture of our Solar System on the big screen. "Look at all the planets. This is our Solar System. It is made up of the eight planets that orbit our star, which we call the Sun. In addition to planets, the Solar System consists of the planets' moons, comets, asteroids, minor planets, dust, and gas," said Dr. Blake.

"The sun is a star?" asked Oni.

"Of course!" said Dr. Blake. "Our sun is an average sized star in the galaxy. It looks much bigger to us because of its distance from our planet. The sun is much closer to earth than all the other stars."

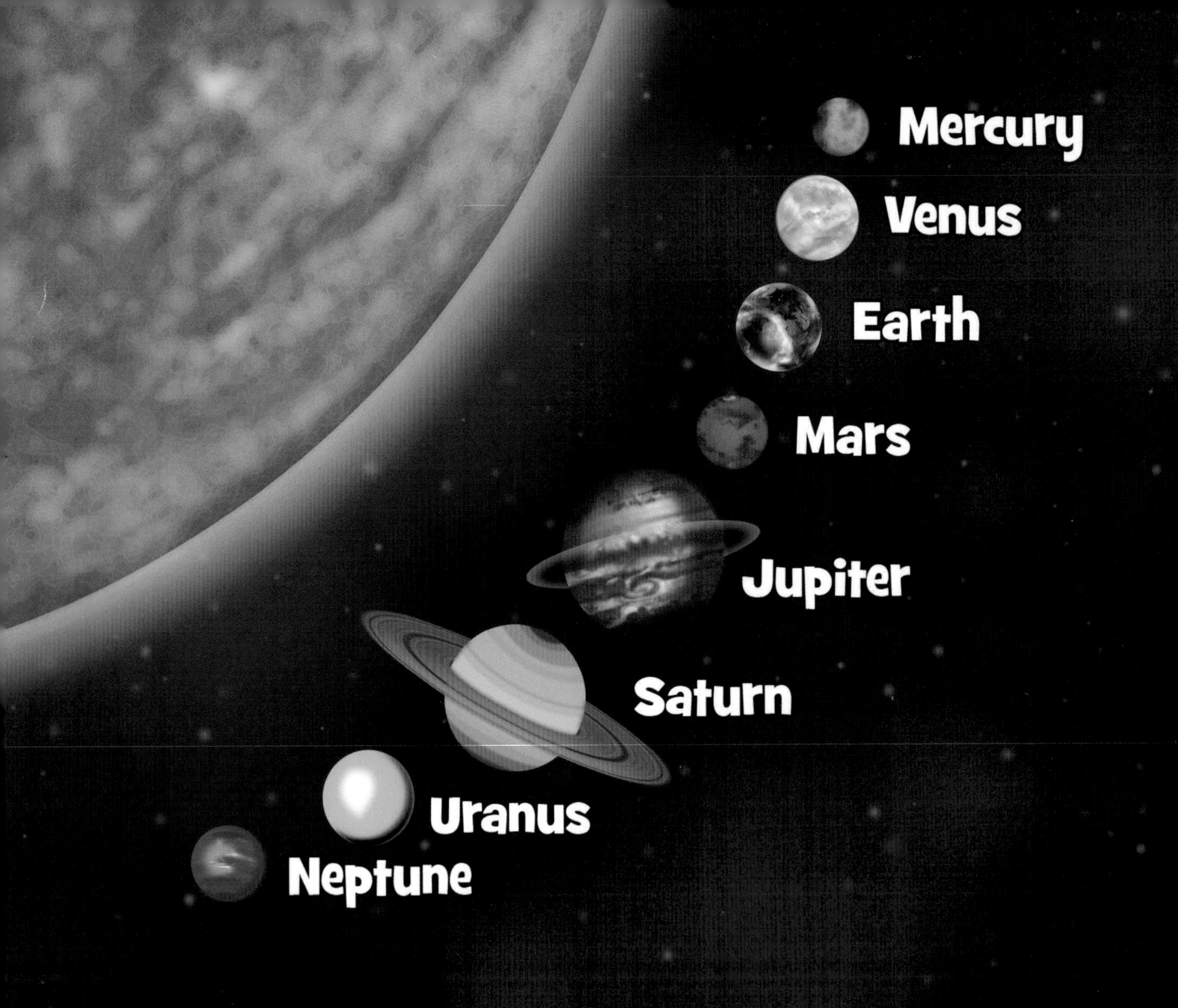

"Here are the names of the eight planets in our Solar System," said Dr. Blake, "Mercury, Venus, Earth, Mars, Jupiter, Saturn, Uranus, and Neptune. One way to remember the planets and their order is to say '*My Very Educated Mother Just Served Us Noodles*'. Each of the first letters will help you to remember the order of the planets."

"Alright kids, we are going to watch a movie about our Solar System," said Dr. Blake. "We will talk about the planets as you see them. Sit back, relax, and enjoy!"

The Alphabet Kids sat way back and stared at the ceiling of the large dome. All of a sudden a big yellow ball with red spots appeared.

"That's the sun!" called Oni. "Our star!"

"That's right," said Dr. Blake, "We can see it clearly because it is on the screen, but when you are looking at the sun outside, you should never stare directly at it."

As the Alphabet Kids were looking at the sun, a round shape appeared on the screen.

"That is the smallest of the eight planets." Dr. Blake said, "It is the closest to the sun and is extremely hot. Does anyone know the name of that planet?"

Yang put up his hand and couldn't wait to answer, "It's Mercury!"

"That is correct," said Dr. Blake.

"Venus is the second planet from the Sun. It is around the same size as our planet Earth, and it is the brightest of all the planets. It is covered in clouds of gas that capture the sunlight, making the planet very hot, even hotter than Mercury," said Dr. Blake. "It has a whitish-blueish color. On some nights, Venus can be seen from Earth without a telescope."

Soon a big light blue colored planet swooped across the ceiling. "Venus! Venus!" said the Alphabet Kids excitedly.

"I know what comes next," said Allegra. "Our planet Earth!"

Soon a planet with land and water appeared on the screen. "Did you know that 80 percent of our earth is water?" said Dr. Blake.

"Wow, it's like a giant swimming pool!" said Umar.

"Mars is the fourth planet from the Sun and is known as the Red Planet," said Dr. Blake. A giant red planet soared across the ceiling.

"It's red just like my shirt!" said Isaac.

"Why is it that color?" asked Yang.

"The rocks and soil have iron in them, making Mars a red color," Dr. Blake responded.

All of a sudden, a giant orange planet appeared on the dome.

"Wow, that is a big planet!" said Allegra with her eyes open wide.

"*Sí, que grande*," said Elena. "Yes, that's big!"

"This is Jupiter, the largest planet in the Solar System,"explained Dr. Blake. "It's more than 1,300 times bigger than our Earth."

Then another planet appeared with many rings around it.

"Saturn!" all the Alphabet Kids said.

"Yes," laughed Dr. Blake. "Saturn is the sixth planet from the Sun and is the second largest planet."

"It's amazing," said Oni.

When a light blue planet appeared Oni called out, "Here comes Uranus!"

"Yes," agreed Dr. Blake, "Uranus is the seventh planet from the Sun and the third largest planet."

"It looks like a huge gumball," said Isaac licking his lips.

Finally, Neptune was shown on the screen and the Alphabet Kids learned that Neptune has eight moons.

"Speaking of moons," said Dr. Blake, "Earth has one moon and we can only see one side of the moon from Earth."

"Being on the moon is very different than being on Earth," Dr. Blake continued, "Twelve astronauts have walked on the moon, so far. They needed air tanks and special suits to help them survive without oxygen and gravity. Our moon is covered with rocks that are about 4.3 billion years old. There are big round holes called craters, which are formed from collisions from either comets, asteroids, and meteoroids."

"What's gravity?" asked Umar.

"Gravity is a force that pulls two objects towards each other. It is what keeps you, and the things around you, on the surface of the earth," said Dr. Blake.

When the lights came on, Dr. Blake said, "I hope you enjoyed the show. Now you can explore the rest of the planetarium to see other wonderful exhibits."

"I want to see a spaceship!" said Yang.

The Alphabet Kids walked around learning more about our Universe. They even got to see a model of a shuttle that travels to outer space.

At the end of the day, Dr. Blake gathered the Alphabet Kids around and said, "Well kids, thanks for coming to the planetarium. I hope you enjoyed the trip and now you know a lot of interesting facts about our Solar System and the planets. Please come back and visit us again."

All of the Alphabet Kids climbed on to the bus. During the ride back to the Alphabet Afterschool Center, they all fell asleep and dreamed about space and their new friend Dr. Blake, the astronomer.